The Juggling Monkeys

QUANTUM

A Quantum Runtime Thought Experiment

by John L Eldred

THE JUGGLING MONKEYS

A Quantum Runtime Thought Experiment

by John L Eldred

Redactions or omissions may or may not be included and any extra or missing information, or changes in the text are necessary…

The weight of fruit…
the weight of knowing.

The Eleven Juggling Monkeys of Quantum Runtime
A Thought Model for Visualizing Entangled System Behavior and
Runtime

Summary

This report presents an expansive and allegorical framework—"The Juggling Monkeys Theorem"—as both a pedagogical tool and operational metaphor to illustrate quantum runtime behavior within computing architectures, especially those using chains and processes. Each Juggling Monkey (JMT) embodies a functional register or runtime variable in the system. Through simulated juggling behavior of quantum fruit analogs, the theorem visually conveys the principles of execution, register, observer-induced collapse, and high-fidelity.

This document integrates logic, functionality, and runtime scheduling with the imaginative use of 11 highly trained metaphorical monkeys. The result is an accessible yet technically faithful guide to the inner workings of advanced quantum runtime mechanisms.

Table of Contents

1. Introduction and Origins

What began as a whimsical joke—"What if a group of juggling monkeys verified quantum output?"—rapidly evolved into a coherent and educational model for understanding quantum runtime behavior. The Juggling Monkeys "Theorem" was born from a need to explain highly complex quantum system behavior using accessible, symbolic operations, while still retaining the operational fidelity of entangled quantum computing systems.

Each monkey symbolizes a runtime component or register. Their fruit represents the runtime state of specific quantum variables. Their behavior—juggling, collapsing, dropping—becomes a metaphorical lens through which runtime behavior, error resolution, and quantum-classical transitions can be intuitively understood.

2. Allegorical Juggling Framework Overview

The allegorical structure of the Juggling Monkeys Theorem is simple in form but deep in metaphorical range. Each monkey is responsible for juggling a specific fruit, chosen for its metaphorical affinity to a quantum variable or runtime role. The juggling action itself symbolizes dynamic, non-collapsed computation. Each cycle of juggling maps to operational slot rotations. If an observer looks at the monkey mid-juggle, the monkey collapses its fruit state into a resolved configuration, mimicking wavefunction collapse.

Crucially, while the metaphor is playful, the framework allows clear analogs to quantum properties such as spin, phase, positional axes, and entanglement lineage. By visualizing these operations through monkeys and their juggling patterns, both novices and experienced quantum theorists can grasp system interrelations that are otherwise mathematically dense.

Key operational assumptions:

- **Each fruit type represents a variable state**
- **Juggling cycles are computation slots**
- **Monkeys only juggle while unobserved (Heisenberg metaphor)**
- **When observed, one fruit lands in hand, others freeze midair (state resolution)**

This paradigm is not merely pedagogical—it creates a narrative-based symbolic language for tracing runtime logic and register interactions.

3. Mapping Monkeys to Quantum Runtime Registers

Each of the eleven monkeys map to one of the eleven key runtime variables in the JMT structure. These include spatial alignment, phase, spin, parity, and temporal markers:

1. **Monkey 1** – XJMT (X-axis presence alignment) – Juggles Oranges

2. **Monkey 2** – YJMT (Y-axis presence alignment) – Juggles Kiwis

3. **Monkey 3** – ZJMT (Z-axis presence alignment) – Juggles Plums

4. **Monkey 4** – ψJMT (Quantum angular phase) – Juggles Grapes

5. **Monkey 5** – sJMT (Spin orientation) – Juggles Spinach Leaves

6. **Monkey 6** – pJMT (Parity information) – Juggles Pineapples

7. **Monkey 7** – ϕJMT (Rotational state) – Juggles Tangerines

8. **Monkey 8** – STΔtJMT (Temporal anchor) – Juggles Time Bananas

9. **Monkey 9** – STAGJMT (tag lineage) – Juggles Mangoes

10. **Monkey 10** – Resolution Sync – Juggles "Clockfruit"

11. **Monkey 11** – Composite Ion State – Juggles the Mega-Merge Fruit (varies per simulation)

Each monkey's juggling behavior simulates state preservation across uncollapsed runtime operation. Their coordinated actions during a cycle represent chain-wide runtime execution. The fruit types encode metaphorical qualities:

- Oranges: Positional consistency

- Kiwis: Multi-spin representation

- Spinach: Spin and fragility

- Time Bananas: Temporal state caching

- Clockfruit: Rotational time registration

- Mangoes and Mega-Merge: Final state assembly and resolution

In runtime simulation, each juggling cycle advances the logical state, and registered values are read only upon "observation," triggering resolution.

4. Fruit as Runtime Variable Carriers

Each fruit in the Juggling Monkeys metaphor was chosen with care, not only for the humorous visual of monkeys tossing them in the air, but for their allegorical strength in representing real quantum runtime variables. This section outlines the symbolic and operational alignment of each fruit type and its corresponding quantum function.

- **Oranges (XJMT)** – Oranges are round, unchanging, and predictable in motion. Their consistent shape and mass symbolize positional certainty along the X-axis. The monkey juggling oranges (Monkey 1) never alters tempo, representing fixed positional stability.

- **Kiwis (YJMT)** – Covered in fine fuzz and naturally imbalanced, Kiwis rotate uniquely midair. This inherent spin variety symbolizes vector wobble or precessional variance in the Y-axis state alignment. Monkey 2 must juggle with intuitive torque correction.

- **Plums (ZJMT)** – Dense and plump, plums fall more directly with less side spin. They represent gravitational certainty and axial consistency, matching Z-axis anchoring. Their juggle is the most stable.

- **Grapes (ψJMT)** – Lightweight, susceptible to "wind" and chaotic bounce, grapes are perfect for angular phase analogs. Their high sensitivity to movement direction and energy transfer represents the subtlety of ψ (wave phase).

- **Spinach Leaves (sJMT)** – With minimal mass and chaotic aerodynamics, spinach leaves are the toughest to juggle. They curl, invert, flap, and reflect spin instability. Monkey 5's expertise with these symbolizes mastery of spin-flip dynamics and collapsed state fidelity.

- **Pineapples (pJMT)** – Irregular, thorny, heavy, and high impact—parity's no joke. These juggle slowly and with risk. Their asymmetry represents systems whose states resist smooth transformation. The pineapple is parity's perfect chaotic emblem.

- **Tangerines (ϕJMT)** – Small, smooth, and rotationally balanced, tangerines can be spun on axis mid-juggle. These map cleanly to ϕ, the runtime's internal angular momentum component.

- **Time Bananas (STΔtJMT)** – The curved arc of a banana naturally represents time travel through non-linearity. The Δt register rides on its orientation—whether tip or stem is leading—guiding temporal anchoring in chain computation.

- **Mangoes (STAGJMT)** – These represent identity. Mangoes are rich, uniquely shaped, and often color-graded. Their juggle tracks state tags and Catch lineage.

- **Clockfruit (JMT Resolution Sync)** – A fictional fruit resembling a hybrid of passionfruit and clock gears, this fruit ticks subtly mid-air, representing runtime ticking alignment. It is used to align the juggling cycles of all monkeys.

- **Mega-Merge Fruit (JMT Composite Build)** – Appears only during simulations. This synthetic fruit morphs into different forms based on runtime variables. It contains DNA of all other fruits and is only juggled by Monkey 11.

Phillip: Acting as
Monkey #11:

5. Observer Effect: Collapse and Resolution

The core of quantum measurement—the observer effect—is beautifully embodied in the Juggling Monkeys framework. While monkeys are unobserved, they juggle flawlessly, maintaining runtime logic in coherent, entangled superposition. But once observed, the juggling ends.

Each monkey reacts in a similar way:

- **One fruit is caught and held**—this represents the resolved value at the time of measurement.

- **Remaining fruits are frozen midair**—mirroring how quantum possibilities "freeze" upon observation.

- **The orientation, spin, and spatial configuration** of each fruit is measured and recorded.

Importantly, the system must recover its computational truth from this resolution moment. Therefore, the juggling output across all monkeys must be interpreted as a simultaneous, interlocked collapse—yielding a full snapshot of system state.

Collapse does not imply error—it implies transition from quantum runtime to classical registration.

To aid understanding, a series of simulations follows to demonstrate these principles in live runtime juggling scenarios.

6. Simulation #1: 3|4 Juggling Operation

In this simulation, we model a simplified logic execution represented by three juggling monkeys across four computation slots (3|4). The monkeys simulate a rotational entanglement structure, where the logical state flows clockwise and collapses on demand.

Setup:

- **Active Monkeys:** Monkey 3 (ZJMT – Plums), Monkey 4 (ψJMT – Grapes), Monkey 6 (pJMT – Pineapples)

- **Cycle Length:** 4 juggling slots

- **Juggling Rule:** Each monkey tosses fruit to the next monkey in the ring after each cycle

Juggling Cycle Log:

1. Cycle 1:
   - M3 tosses Plum to M4
   - M4 tosses Grape to M6
   - M6 tosses Pineapple to M3

2. Cycle 2:
   - M3 tosses Pineapple to M4
   - M4 tosses Plum to M6
   - M6 tosses Grape to M3

3. Cycle 3:
   - M3 tosses Grape to M4

- o   M4 tosses Pineapple to M6

- o   M6 tosses Plum to M3

4. Observation Triggered:

- o   M3 catches Plum

- o   M4 catches Grape

- o   M6 catches Pineapple

Analysis:

- The final collapsed values mirror the original configuration, signaling a successful logic cycle with no decoherence.

- Fruit paths emulate data flow in Chains.

- The slot-based interaction preserves register timing and entangled integrity.

Analyst Commentary:

"This small-scale juggling simulation successfully demonstrates runtime flow-through and symmetry. Monkey 6's pineapple handling was especially notable due to parity's instability, but resolved cleanly. Juggling rotation as quantum routing? Monkey business or genius."

7. Simulation #2: (Full JMT Build)

In this simulation, the monkeys execute a full (JMT) operation. A Scrambled State is passed through JMT routines to construct/reconstruct a logically equivalent State and some may presently be thinking about Quantum Error Correction or (QEC). All eleven monkeys participate.

Setup:

- **Source:** Scrambled phase, spin, and states
- **Target:** Stable reconstruction
- **Trigger:** Juggling begins once observer triggers recovery signal

Fruit Initial State:

- Each monkey begins with inverted or chaotic form of their fruit (e.g., upside-down banana, oblong plum, off-phase grape)

Cycle 1 – Correction Initiation:

- Monkeys spin fruit back to canonical orientation
- Monkey 9 (Mango) flashes green, indicating correct lineage tag
- Monkey 4 (ψ – Grapes) synchronizes phase with Monkey 11

Cycle 2 – Runtime Alignment:

- Clockfruit (Monkey 10) ticks align with Time Bananas (Monkey 8)
- Monkeys 2 and 3 resolve axis overlaps via coordinated tosses

- Spinach (Monkey 5) remains unstable, flapping wildly

Cycle 3 – Final Juggling Pass:

- All fruits show uniform trajectory, velocity, and juggling height
- Observer-triggered collapse initiated

Observation Result:

- 11 monkeys catch fruit simultaneously
- All fruits exhibit canonical properties: corrected spin, aligned phase, matching axis

Outcome:

"Successful full register construct/reconstruction from a scrambled state. Even Monkey #5 nailed the spinach. The operation indicates high-fidelity resolution and confident parity."

Monkey #5 is a very talented juggler.

8. Pedagogical Benefits and Interpretability for Reviewers

The Juggling Monkeys Theorem provides an immediate and engaging entry point for individuals unfamiliar with quantum runtime systems—particularly those tasked with reviewing or validating patents, theoretical frameworks, or software execution logic related to quantum computing platforms.

Key Benefits:

- **Conceptual Clarity:** By converting abstract quantum states into physical metaphors (e.g., grapes for wave phase), the model demystifies runtime behavior.

- **Visualization of Coherence:** Slot-based juggling metaphorically aligns with coherent logic cycles and runtime stability, showing how operations propagate over time.

- **Observer-Collapse Analogy:** The act of stopping the juggle illustrates wavefunction collapse, data resolution, and measurement interference with exceptional clarity.

- **Chain Resolution Understanding:** When reviewers see 11 monkeys resolve fruit states simultaneously, the unity of quantum collapse is viscerally felt.

This interpretive layer is not merely cute—it can be foundational for:

- Training new quantum developers

- Teaching students and interns

- Illustrating runtime behavior in educational videos

- Communicating patent mechanisms without excessive mathematical scaffolding

If visualized or animated, the Juggling Monkeys metaphor could serve as a short-form explainer, quickly familiarizing legal or academic audiences with runtime principles before deep technical review.

Ultimately, the theorem functions as both a scientific pedagogical aid and a creative diagnostic model for runtime verification.

9. Fidelity Assessment of Juggling-Based Output

Assessing the fidelity of computational output from the Juggling Monkeys metaphor requires framing expectations: Are we aiming for whimsical allegory or diagnostic utility? Surprisingly, the answer may be both.

Fidelity Indicators:

- **Slot-Based Accuracy**: Each monkey's juggling rhythm mimics discrete logic slots in execution. When monkeys juggle at synchronized cadence, runtime harmony is achieved.

- **Variable Preservation**: Fruit states held upon observation retain metaphorical alignment with expected logic values in a simulated runtime.

- **System-Wide Collapse Consistency**: When observed, all monkeys collapse in unison, yielding a coherent runtime resolution snapshot—analogous to system-wide classical readout.

- **Error Signaling via Drop Events**: Fruit drops, spins, and mid-juggle anomalies correlate to error propagation, parity breaks, or decoherence events.

- **Observer Bias Simulation**: Different angles of observation (e.g., phase-locked vs. free-observer) yield different resolved states, introducing a realistic approximation of measurement-dependence.

Use Cases:

- **High-Fidelity Runtime Diagnostics**: Simulations where juggling fidelity is maintained throughout may parallel successful JMT resolution.

- **Fault Injection Modeling**: Dropped fruit can simulate external decoherence, ion loss, or misaligned rotational state recovery.

- **Behavioral Visualization**: The intuitive grasp of runtime fidelity becomes evident when simulating multiple juggling rounds.

Limitations:

- **Symbolic Saturation**: The metaphor becomes abstract beyond 11 variables. Scaling the model requires care or splitting monkeys into performance groups (e.g., Core Runtime vs. Spectral Observers).

- **No Actual Quantum Mechanics**: The theorem does not simulate wavefunctions, entanglement strength, or probabilistic resolution. It mirrors conceptual operations only.

Yet, within this scope, the theorem excels. Its fidelity is pedagogically optimized rather than numerically calibrated. The patterns it illustrates—runtime harmony, coherent collapse, error signaling, and observer sensitivity—faithfully reflect operational truths in quantum platforms.

In this light, the Juggling Monkeys model achieves its core goal: to be useful, memorable, technically evocative, and insight-rich.

10. Comparisons to Real Quantum Runtime

To evaluate the utility of the Juggling Monkeys metaphor, it must be weighed against the actual operational mechanics of a quantum runtime environment. Despite the whimsical tone, the analogy maps cleanly to several critical behaviors in the runtime system.

Computation Alignment

- **Juggling Rhythm = Slot Timing**: The cadence and phase of a monkey's juggling represent slot-based execution in a chain.

- **Fruit Type = Runtime Register**: Each fruit maps directly to a specific runtime variable (spin, parity, phase, etc.) as described in the JMT framework.

- **Cross-Monkey Coordination = Entangled Instruction Synchronization**: When multiple monkeys juggle in coordination, they reflect the entanglement and dependency resolution across quantum instruction packets.

Post Catch via Juggling

- **Observer-Collapse Mirrors Measurement**: The point when an observer looks—and one fruit is caught—accurately portrays runtime collapse logic.

- **State = Chaotic Juggling Phase**: Chaotic juggling at the beginning of Simulation #2 parallels the disordered.

- **Resolution = Canonical Fruit Arrangement**: Once juggling stabilizes, the outcome matches a successful Chain using the JMT.

Limitations and Enhancements

- **Multi- Extension**: Real runtimes may operate with layered or forks—beyond what 11 monkeys can illustrate without expansion.

- **Monkeys as Runtime Diagnostics**: The model may overemphasize harmony; real systems embrace stochastic and probabilistic branching that would require jittery or improvisational juggling.

Nonetheless, the comparison confirms that the metaphor has legs. In performance and interpretability, the Juggling Monkeys embody a significant subset of runtime behaviors, including timing, entanglement, collapse, error, recovery, and identity stabilization. When used alongside technical documentation, the monkeys offer vivid reinforcement of core runtime ideas within quantum computation environments.

11. Feasibility, Fault Tolerance, and Diagnostic Potential

The value of the Juggling Monkeys Theorem expands considerably when examined through the lens of engineering feasibility and fault resilience.

Feasibility

While no actual quantum computer employs simian logic handlers, the metaphor's architectural correspondence is robust. Slot-based juggling maps to sequenced quantum gates. Spin states correlate with rotational fruits. Collapse-to-hand mirrors collapse-to-state. From an instructional and planning perspective, the allegory is fully feasible.

- **Fidelity**: Juggling cadence mimics quantum gate timing.

- **JMT Variables**: Each fruit represents a clean metaphor for quantum state alignment.

- **Observation Response**: Modeled collapse aligns with quantum readout behavior.

Fault Tolerance

Juggling Monkeys naturally express fault domains:

- **Dropped Fruit**: Simulates decoherence or failed measurement.

- **Errant Spin**: Represents spin-orbit or parity mismatch.

- **Desync**: Reflects timing violations in gate sequences.

In particular, the analogy captures one of quantum computation's hardest problems—cross-system error propagation—and encodes it in visual, interpretable terms.

- **Group Collapse Divergence**: If all monkeys collapse except one, it mirrors partial decoherence or asynchronous job halt.

- **Juggling Recovery**: When monkeys resume from a stable position, it simulates QEC-assisted runtime reinitialization.

Diagnostic Utility

Engineers or reviewers can use the model to map runtime anomalies:

- **Determine Fault Origin**: Track which monkey (register) dropped fruit.

- **Evaluate Collapse Symmetry**: Was system-wide collapse ordered or stochastic?

- **Project Recovery Feasibility**: Does the state permit feasible regeneration?

In environments where quantum behavior must be monitored in real time—such as defense or space-grade quantum platforms—the visualization framework may guide:

- Runtime monitoring dashboards

- Pre-job simulation stress testing

- Dynamic instruction re-routing alerts

Ultimately, the Juggling Monkeys serve as more than entertainment. They form a diagnostic lens for runtime analysis, helping trace logic flow, register health, collapse harmony, and recoverability.

12. Addendum Role

Given the unique visual, diagnostic, and educational strengths of the Juggling Monkeys Theorem, this conceptual framework is formally noted as being associated with patent filings.

While not itself a claim of utility or novelty in the statutory sense, the Juggling Monkeys allegory operates as a supplemental comprehension tool. By mapping quantum logic behavior, runtime failure states, and operations to distinct metaphorical elements, this model accelerates understanding for patent examiners, technical analysts, and interdisciplinary stakeholders.

Recommended Integration Method

"A visual and conceptual theorem—The Eleven Juggling Monkeys of Quantum Runtime—has been developed to aid in the interpretation and communication of a system's operational and logical behavior. A full pedagogical report may be made available upon request."

This statement ensures that the theorem is officially referenced without diluting the focus of the core patent language.

Benefits for Review

- **Accelerates Comprehension**: Helps reviewers rapidly internalize runtime logic.

- **Reveals Edge Case Behavior**: Fruit drops, juggling pauses, and collapse irregularities provide conceptual tools for edge conditions.

- **Supports Interdisciplinary Review**: Teams without deep quantum background (e.g., software engineers or educators) gain interpretability.

In summary, inclusion of the theorem as an optional educational supplement in documents supports faster, clearer, and more holistic evaluation of the advanced architectural sophistication and functional elegance.

13. Future Expansions: Q-Circus and Runtime Visualization Studios

The Juggling Monkeys Theorem opens a new conceptual frontier: turning runtime quantum logic into live or animated visualization experiences. Under the working name **Q-Circus**, proposed, is a transdisciplinary expansion of the theorem into interactive runtime simulation environments.

13.1 Q-Circus: Animated Runtime Studio

A digital environment where juggling monkeys simulate real quantum operations:

- **Live Quantum Job Playback**: Translate actual quantum job logs into juggling routines.

- **Runtime Failure Theater**: Visually play out edge case collapses or decoherence cascades.

- **Register Behavior Spectacle**: Show each monkey juggling specific fruits to interpret logic operations.

13.2 Runtime Visualization Studio

Inspired by cockpit simulators and network topology viewers, this studio would allow:

- **Interactive Runtime Walkthroughs**: Trace computation and JMT through Monkey choreography.

- **Developer Mode**: Inject runtime anomalies to observe juggling error manifestation.

- **Observer Overlay**: Introduce artificial "observers" and trigger collapse.

13.3 Broader Educational Tools

With additional funding or support, the Q-Circus model could evolve into:

- **Museum Exhibits**

- **STEM Outreach Kits**

- **AI-Coached Juggling Simulations for Training New Quantum Engineers**

By combining quantum logic with storytelling and kinetic visualization, Q-Circus becomes an educational powerhouse—simplifying complexity without trivializing it.

14. Conclusion and Next Steps

The Juggling Monkeys Theorem began as a light-hearted thought model, yet has grown into a rich and multifaceted allegory for describing, testing, and even diagnosing complex quantum runtime phenomena. From slot-based logic to fidelity, the metaphor proves not only educational but insight-generating. Its metaphorical register-fidelity aligns tightly with the operational roles of real quantum chains under scheduling frameworks.

Through extensive simulations—both playful and technically grounded—this report has demonstrated the feasibility of mapping runtime logic, state collapses, and processes into kinetic, interpretable visual metaphors. In turn, these simulations offer a mental model that is immediately graspable by audiences ranging from quantum engineers to professors, reviewers and students.

Immediate Opportunities:

- Integrate the theorem as a reference in related documents.

- Publish this report as a supplementary appendix or public educational white paper.

- Explore development of Q-Circus runtime visualizer prototypes.

Mid-Term Projects:

- Design interactive runtime simulators based on juggling behavior.

- Develop a toolkit for identifying failure types through monkey collapse profiles.

- Expand the fruit-type logic model for advanced quantum register encoding.

Long-Term Vision:

- Create museum-grade educational installations featuring Q-Circus.

- Align monkey choreography with real runtime logs from quantum systems.

- Introduce AI-driven monkey trainers for auto-simulation of runtime stress cases.

In short, this may only be the beginning. The quantum juggling monkeys—through their chaos, coordination, and collapse—may well become the ambassadors of complex runtime understanding for a new generation of engineers, educators, and observers.

Appendix A: Individual Monkey Profiles

Monkey 1: XJMT Juggler

Fruit: Orange (stable, identical)
Register Role: X-axis presence alignment.
Juggling Behavior: Predictable loops in X-plane space. Demonstrates high-fidelity retention of horizontal logic. Responds precisely to observer collapse with minimal variance.

Monkey 2: YJMT Juggler

Fruit: Kiwi (textured, varied spin)
Register Role: Y-axis presence alignment.
Juggling Behavior: Slightly chaotic but maintains vertical lane consistency. Demonstrates runtime phase drift compensation.

Monkey 3: ZJMT Juggler

Fruit: Pineapple chunks
Register Role: Z-axis presence alignment.
Juggling Behavior: High-apex tosses in depth axis; good for vertical displacement register reconstruction in entanglement layer shifts.

Monkey 4: ψJMT Juggler

Fruit: Grapes
Register Role: Quantum angular phase.
Juggling Behavior: Rapidly spinning, multiorbital juggling. Critical for assessing angular entanglement state fidelity and phase sync.

Monkey 5: sJMT Juggler

Fruit: Spinach leaves
Register Role: Spin orientation.
Juggling Behavior: Flat, fragile, feather-like tosses that respond sensitively to micro-adjustments. Key for quantum spin vector integrity.

Monkey 6: pJMT Juggler

Fruit: Bananas
Register Role: Parity.
Juggling Behavior: Always throws fruit in mirrored pairs. Even/odd transitions immediately evident upon observer collapse.

Monkey 7: φJMT Juggler

Fruit: Cherries with stems
Register Role: Rotational state.
Juggling Behavior: Stem spin is tracked precisely mid-air. Registers angular torsion and twist encoding.

Monkey 8: STΔtJMT-A (Stabilizer)

Fruit: Mango slices
Register Role: Temporal stabilizer.
Juggling Behavior: Consistent rhythm, anchoring time dilation and logic delay mapping across chain timing windows.

Monkey 9: STΔtJMT-B (Δt Recorder)

Fruit: Pomegranate seeds
Register Role: Timestamp registration.
Juggling Behavior: Rapid micro-tosses encoding Δt offsets. High refresh rate. Essential for timeframe registration.

Monkey 10: STAGJMT-1

Fruit: Dragonfruit
Register Role: tag encoding.
Juggling Behavior: Complex swirling paths with distinct identifiers. Key for post-juggling chain decoding.

Monkey 11: STAGJMT-2

Fruit: Starfruit
Register Role: origin tracer and runtime label keeper.
Juggling Behavior: Star-shaped tosses in alternating planar loops, encoding system origin and layer transitions.

Appendix B: Cross-Monkey Entanglement Interference Cases

This appendix documents known or simulated cases where the entanglement state of one Juggling Monkey influences the juggling fidelity of another. These events model interference, entanglement leakage, state bleed, or unintended coupling between registers in runtime operations.

Case B.1: Spin-Induced Phase Distortion

- **Monkeys Involved:** Monkey 5 (sJMT) and Monkey 4 (ψJMT)

- **Observation:** Irregular fluttering in spinach leaf trajectories caused angular misalignment in the grape orbitals.

- **Interpretation:** Cross-register entanglement effects are possible when spin orientation destabilizes angular phase encoding.

Case B.2: Timestamp Drift with Y-Axis Overlap

- **Monkeys Involved:** Monkey 2 (YJMT) and Monkey 9 (STΔtJMT-B)

- **Observation:** Varied kiwi spin introduced timing irregularities that led to premature Δt seed drops.

- **Interpretation:** Spatial register behavior (Y-axis) can influence micro-timestamp alignment during high-speed juggling.

Case B.3: Origin Label Swapping in Layer Transition

- **Monkeys Involved:** Monkey 11 (STAGC-**JMT**) and Monkey 3 (Z**JMT**)

- **Observation:** Depth tosses of pineapple chunks coincided with altered planar starfruit paths.

- **Interpretation:** Runtime origin labels may bleed into dimensional registers during vertical layer transitions.

Case B.4: Parity Jumps Affecting Stabilizer Rhythm

- **Monkeys Involved:** Monkey 6 (p**JMT**) and Monkey 8 (STΔt**JMT**-A)

- **Observation:** Mirrored banana throws caused phase gaps in mango slice rhythm.

- **Interpretation:** Temporal stabilization logic is vulnerable to sudden parity oscillations.

These cases highlight the need for runtime shielding protocols to suppress register-level cross-talk and maintain fidelity under stress or collapse events.

Appendix C: Reconstructed from Juggling Output

This appendix (re)constructs a **JMT** sample using only the output states derived from a controlled simulation of the 11 Juggling Monkeys. Each monkey's juggling behavior was captured post-collapse to determine the terminal state of their fruit and associated register. The goal was to engineer a correct metaphorical register using values encoded in monkey performance.

Reconstructed JMT Table (Abbreviated Sample)

Register	Monkey #	Fruit Type	Final Position (Juggle End)	Interpreted Quantum Value
XJMT	1	Orange	Left hand catch	X-axis: $+1$
YJMT	2	Kiwi	In-air spin (counter)	Y-axis: -1
ZJMT	3	Pineapple	Dropped center mass	Z-axis: 0
ψJMT	4	Grapes	Splayed across arc	Phase offset: $\pi/4$
sJMT	5	Spinach	Fluttered, landed edgewise	Spin: Up

Register	Monkey #	Fruit Type	Final Position (Juggle End)	Interpreted Quantum Value
p**JMT**	6	Bananas	Even pair caught right hand	Parity: Even
φ**JMT**	7	Cherry	Twisting, stem aligned N-E	Rotation: 90° CW
STΔt**JMT**-A	8	Mango slices	Rhythm matched base beat	Stabilizer: Active
STΔt**JMT**-B	9	Pomegranate	Seed spiral out @ 0.002s diff	Δt: +0.002s
STAG**JMT**-1	10	Dragonfruit	Ident trace looped	Tag: Catch-17
STAG**JMT**-2	11	Starfruit	Patterned flip w/ Z crossover	Layer: 4

JMT Integrity Validation:

Using this (re)construction, the results showed 96.8% alignment with expected post-logic. Deviations occurred in rotational torsion inference (φ**JMT**) and timestamp drift (Δt**JMT**-B), both of which fell within acceptable error margins for (re)construction.

Appendix C: Commentary:

- The juggling metaphor preserves slot-by-slot register fidelity.

- Observer collapse was used to simulate measurement interruption.

- Monkey interactions (from Appendix B) were logged to identify entanglement leaks.

This demonstration confirms the practical interpretability of monkey-based quantum modeling and its potential use in trace diagnostics or pedagogical environments.

Appendix D: Runtime Logs in Juggling Format (Example Traces)

This appendix presents sample runtime logs translated into juggling metaphor format to showcase the dynamics of active logic slot transitions, fruit-hand offloads, and observer-induced resolutions. These traces mirror live runtime scenarios in highlighting the expressive power of this allegorical framework.

Sample Runtime Trace 1: Computation Cycle (3|4)

Monkey 1 (Orange) — Slot Pattern: Left → Arc → Right → Drop → Left (Catch)
Monkey 2 (Kiwi) — Slot Pattern: Spin CW → Float → Spin CCW → Caught Mid-Air

Logic Interpretation:

- Stable trajectory detected on Register XJMT

- Active logic flip from kiwi spin pattern yields alternate Y-alignment

Sample Runtime Trace 2: Full JMT Register Collapse Post-Catch

Monkey 4 (Grapes) — Scatter detected → Arc dispersion
Monkey 5 (Spinach) — Spin mid-juggle, flutter landed North
Monkey 10 (Dragonfruit) — Trace echoed back after bounce off trapeze net (tag integrity verified)

Runtime Notes:

- successful; 98.2% registered

- Starfruit (Monkey 11) flipped with mirrored Z-line crossover, triggering successful audit

Diagnostic Utility:

- The logs support entanglement causality analysis

- Deviant fruit trajectories mark slippage points in routines

- Trace entries enable retro-diagnostics using fruit-flight timing

These translated logs serve as both humorous and technically coherent models for analyzing runtime integrity, fruit-based encoding fidelity, and operational synchrony in complex quantum systems.

Monkey #11: Wendell showing off the computational output.

Notes from the Author: The Juggling Monkeys Theorem

The theorem was to be this. I would propose to use Juggling Monkeys to verify the computational output of a Quantum Computer by AI.

But by the time I got to step 3… it got real… So… here is how it started…

Juggling Monkey #1 can juggle 3 Oranges.

The 3 oranges are all exactly the same size, weight, texture and they never spin as for some reason they are all very stable oranges.

Juggling Monkey #2 can juggle 4 Kiwis.

Every kiwi is "a like" except that each one always has a different type of spin to it.

The monkeys can only juggle their fruit when no one is looking.

If anyone ever looks at the monkeys, for some reason, they always drop the fruit and the fruit lands gently and perfectly as they were while juggled in mid-air, except for one fruit, which is always in the monkey's "catching hand".

OK… So, what started as a "Theorem" to test AI, turned into a thought experiment. Mathematicians, please don't hold the use of the term "Theorem" throughout the text against me.

Again…

Redactions or omissions may or may not have been included and any
extra or missing information, or changes in the text were necessary…

The Juggling Monkeys offer their Deepest Apologies. Have a banana…

Marci

Coming Soon:

How I Trade NQ Futures: A Conversational Compendium

A Practical Guide to Futures, Flow, and Finding Your Edge

Written by the same curious mind behind *The Juggling Monkeys*, this book leaves the circus tent and dives deep into live markets — revealing how structure, discipline, and signal clarity can tame even the wildest price action.

From runtime collapse to order flow chaos... it all connects.

ψJMT

Jed wonders,.. "Maybe wave functions are connected via dark matter or dark energy?"

Featuring:

Monkey #1 Toko

Monkey #2 Happy

Monkey #3 Gracie

Monkey #4 Jed

Monkey #5 James

Monkey #6 Joe

Monkey #7 Mac

Monkey #8 Marco

Monkey #9 Marci

Monkey #10 Phillip

Monkey #11 Wendell

www.ingramcontent.com/pod-product-compliance
Lightning Source LLC
Chambersburg PA
CBHW060038050426

42448CB00012B/3063